The History of Electromagnetism

By **Wasiful Alam**

Alpona Publication
Praha, Czech Republic
October, 2024

Contents

1. Ancient Observations and Early Descriptions 5
   - 1.1 600 BC -Thales of Miletus 5
   - 1.2 Ancient Indus Writings 5
   - 1.3 Ancient Chinese Writings 5
   - 1.4 Earliest use of Compass 6
   - 1.5 The Book of Nature (Taoism) 6
   - 1.6 Theophrastus (371–287 BC) 6
   - 1.7 Pliny the Elder (AD 23–79) 6
   - 1.8 Islamic Golden Age 6
2. 17th Century Renaissance Discoveries 6
   - 2.1 1600 - William Gilbert 6
   - 2.2 1620 - Francis Bacon 8
   - 2.3 1646 - Sir Thomas Browne 8
   - 2.4 1660s - Robert Boyle 9
   - 2.5 1670s - Otto von Guericke 10
   - 2.6 1680s - Isaac Newton 12
   - 2.7 Summary of the 17th Century Contributions 13
3. 18th century 14
   - 3.1 1729 - Stephen Gray: 14
   - 3.2 1733 - Charles François de Cisternay du Fay 15
   - 3.3 1745-1746 Ewald Georg von Kleist and Pieter van Musschenbroek 17
   - 3.4 1752 - Benjamin Franklin 18
   - 3.5 1770s - Joseph Priestley 19
   - 3.6 1785 - Charles-Augustin de Coulomb 21
   - 3.7 1791 - Luigi Galvani 22
   - 3.8 Summary of the 18th Century Contributions 24
4. 19th Century: The Unification of Electricity and Magnetism 24
   - 4.1 1800 - Alessandro Volta 24
   - 4.2 1820 - Hans Christian Ørsted 26
   - 4.3 1820 - André-Marie Ampère 27
   - 4.4 1831 - Michael Faraday 28
   - 4.5 1864 - James Clerk Maxwell 30
5. Late 19th Century to Early 20th Century: Technological Applications 31
   - 5.1 1887 - Heinrich Hertz 31
   - 5.2 1888 - Nikola Tesla: 33

	5.3	1901 - Guglielmo Marconi:	34
	5.4	Summary of the 19th Century Contributions	36
6	20th Century: Quantum Electrodynamics and Relativity		36
	6.1	1900 - Max Planck	36
	6.2	1905 - Albert Einstein	37
	6.3	1920s - Development of Quantum Electrodynamics (QED)	37
	6.4	1930s - Advances in Particle Physics	37
	6.5	1940s - Development of Microwave Technology	38
	6.6	1950s - Advancements in Electronics	38
	6.7	1960s - Quantum Field Theory and Unification	38
	6.8	1970s - The Standard Model and Experimental Confirmation	38
	6.9	1980s - Advancements in Telecommunications	39
	6.10	1990s - Emergence of the Internet and Modern Computing	39
	6.11	Summary of the 20th Century Contributions	39
7	21st Century		40
	7.1	Terahertz Technology Development (2000s)	40
	7.2	Quantum Computing Experiments (2000s-Present)	40
	7.3	Metamaterials Research (2000s-Present)	40
	7.4	Quantum Communication Experiments (2010s-Present)	40
	7.5	Advanced Imaging Techniques (2010s-Present)	40
	7.6	Wireless Power Transfer (2010s-Present)	40
	7.7	Summary of 21st Century Contribution	41
8	Conclusion		41

1 Ancient Observations and Early Descriptions

The natural observations exceed way beyond any historical record. But we can still find some natural evolution that harness the power of electricity and motivated early humans to learn more. The electric eel can produce electric discharges up to 350 volts, which it uses for hunting and self-defense. They are exclusively found in the freshwater habitats of South America, specifically in the Amazon and Orinoco River basins. Electric Rays are found in coastal waters of Africa and Asia. They can generate electric discharges to stun prey and defend against predators. The aboriginal people were well aware of these creatures, but did not have any understanding of how they work.

People who used to live near mountains containing magnetite or who used to mine iron knew about stones that attract.

Thunderstorms were really common near costal regions. Animals used to get burned by thunder fall. In places with dry weather, spark caused by friction from woolen or cotton wearings were also known.

Even primordial humans knew that rubbing two metallic stones will cause a spark. They were not sure about the reasons of these fundamental phenomena which we still observe in our daily life.

As Greeks used to collect all the documents and kept them protected, these documents are considered the oldest source. We begin the historical journey from one of the Greek philosopher.

1.1 600 BC -Thales of Miletus

Thales of Miletus in Ancient Greece observed that rubbing amber with fur would attract small objects like feathers. This is one of the earliest recorded observations of static electricity. The Greek word for amber, "elektron," is the root of the modern word "electricity."

1.2 Ancient Indus Writings

The ancient Indus valley text *Atharvaveda* (c. 1000 BC) mentions the attractive property of certain stones. In the text, there are references to iron and its relationship with magnetic stones. Later, the Indian philosopher and mathematician Bhaskara II (1114–1185 AD) provided more systematic studies on magnetism and its effects.

1.3 Ancient Chinese Writings

Ancient Chinese texts from the Warring States period (475–221 BC) describe the use of magnetic compasses. In the book *The Book of the Devil Valley Master*, there are references to the south-pointing spoon made of lodestone, which was used for directional purposes. The Chinese philosopher Shen Kuo (1031–1095 AD) later wrote about the properties of magnetic needles in his work *Dream Pool Essays*, describing their use in navigation.

1.4 Earliest use of Compass

The first compass is believed to have been invented during the Han Dynasty in China, around the 2nd century BC. Early Chinese compasses were not used for navigation initially but were instead employed for geomancy and divination purposes, aligning with the principles of feng shui.

These early compasses were made from a naturally occurring magnetic ore known as lodestone, which was shaped into a spoon and placed on a smooth surface, often a bronze plate. The handle of the spoon would point south, serving as a directional tool.

It wasn't until the Song Dynasty (11th–12th century AD) that the compass was adapted for navigation, significantly impacting maritime travel and exploration. The technology spread to the Arab world and Europe, where it played a crucial role in the Age of Exploration.

1.5 The Book of Nature (Taoism)

Ancient Chinese texts, especially within the framework of Taoist philosophy, often hinted at concepts of energy and natural forces. While these texts did not explicitly discuss electricity, they explored interactions between objects, which could be seen as an early analogy to later scientific principles.

1.6 Theophrastus (371–287 BC)

Theophrastus, a student of Aristotle, wrote extensively on the properties of minerals. In his work *On Stones*, he described the attraction between certain stones, like magnetite, and iron, contributing to early knowledge of magnetism and the physical properties of minerals.

1.7 Pliny the Elder (AD 23–79)

Pliny the Elder, in his encyclopedic work *Natural History*, provided one of the earliest detailed accounts of minerals and stones, including the properties of magnetite (or lodestone). He observed that magnetite could attract iron.

1.8 Islamic Golden Age

During the Islamic Golden Age, scholars such as Al-Khwarizmi and Ibn Sina (Avicenna) made significant contributions to the understanding of magnetism and the properties of certain stones. Al-Khwarizmi's work on mechanics included observations about magnetic behavior and interactions.

2 17th Century Renaissance Discoveries

2.1 1600 - William Gilbert

William Gilbert(1544-1603), an English physician and scientist.Gilbert's work marked the beginning of the systematic study of magnetism and static electricity. He distinguished between magnetic phenomena and static electrical effects, which was a crucial step towards understanding electromagnetism. He coined the term "electricus" to describe materials that could be electrified by rubbing. He is often regarded as the father of electrical engineering and magnetism.

Work: *De Magnete, Magneticisque Corporibus, et de Magno Magnete Tellure*

- **Contributions:**

    - **Publication of *De Magnete*:** On the Magnet, Magnetic Bodies, and the Great Magnet of the Earth): Gilbert's seminal work published in 1600, was the first major scientific treatise on magnetism and electricity. This book laid the foundation for the study of electromagnetism and influenced scientists for centuries.

    - **Systematic Study of Magnetism:** *De Magnete* was a comprehensive study based on empirical observations and experiments. Gilbert meticulously documented his experiments with magnets, challenging prevailing theories and misconceptions about magnetism.

    - **Magnetism vs. Static Electricity:** Gilbert was the first to distinguish clearly between magnetism and static electricity. He observed that materials like amber (when rubbed) attract light objects due to what we now call static electricity, whereas lodestones naturally exhibit magnetic properties.

    - **Term "Electricus":** Gilbert coined the term *electricus* to describe substances that attract lightweight objects after being rubbed. This term is the origin of the modern word "electricity."

    - **Earth's Magnetic Field:** Gilbert proposed that the Earth itself is a giant magnet, explaining the behavior of compasses. He demonstrated that a spherical magnet could simulate the Earth's magnetic field and showed how this model could explain the orientation of compass needles.

    - **Terrella Experiments:** Gilbert used a small magnetized sphere, called a *terrella* (little Earth), to model the Earth's magnetic properties. His experiments with the terrella provided convincing evidence for his theory of the Earth's magnetism.

    - **Rejection of Aristotelian Science:** Gilbert's work marked a departure from the Aristotelian view that had dominated natural philosophy for centuries. Instead of relying on speculative reasoning, Gilbert emphasized experimentation and observation.

Influence on Later Work: William Gilbert's contributions to the study of magnetism and electricity were foundational. By distinguishing between magnetism and static electricity and proposing the revolutionary idea that the Earth itself is a magnet, Gilbert laid critical groundwork for future scientists. His work in *De Magnete* not only advanced the understanding of magnetic and electric phenomena but also set a new standard for scientific inquiry based on experimentation and observation.

Gilbert's empirical approach and his emphasis on the importance of experimental evidence influenced later scientists, including Galileo Galilei and Johannes Kepler, helping to usher in the Scientific Revolution. Gilbert's ideas and methods would pave the way for subsequent discoveries in electromagnetism, influencing figures such as Benjamin Franklin, Hans Christian Ørsted, André-Marie Ampère, and Michael Faraday. His emphasis on the importance of experimentation in understanding natural phenomena helped shape the development of modern science.

> "The magnet attracts iron. This power is due to the peculiar nature of the magnet, which possesses an inherent virtue, capable of influencing the iron with its force." — *De Magnete*, Book 1

2.2 1620 - Francis Bacon

Francis Bacon (1561–1626) was an English philosopher, statesman, and essayist who is often regarded as the father of empiricism. His works laid the foundation for the modern scientific method, emphasizing the importance of empirical research and inductive reasoning over the traditional Aristotelian approach.

- **Work** *The Advancement of Learning* (1605)

  Novum Organum (1620)

- **Contribution**

  Scientific Method: Bacon's advocacy for empirical research and the scientific method provided a philosophical foundation for experimental science. His approach emphasized observation and experimentation over speculation, which influenced how future scientists would approach the study of electricity and magnetism.

> "Those experiments be not only esteemed which have an immediate and present use, but those principally which are of most universal consequence for invention of other experiments, and those which give more light to the invention of causes; for the invention of the mariner's needle, which giveth the direction, is of no less benefit for navigation than the invention of the sails, which give the motion. "
> — *Sir Francis Bacon, The Advancement of Learning*

2.3 1646 - Sir Thomas Browne

Browne(1605–1682) a polymath, physician, and writer expanded on Gilbert's work in his book *Pseudodoxia Epidemica*, which described various electrical phenomena and helped popularize Gilbert's ideas. His work embodies the spirit of early modern science, as he emphasizes the importance of empirical observation and experimentation over reliance on tradition, authority, or hearsay. He often refers to contemporary scientific discoveries and advancements to support his arguments. in Book II, Chapter 3 of Pseudodoxia Epidemica, Browne discusses the attractive power of materials like amber, diamond, and jet. He explains that when rubbed, these materials exhibit an attractive force on other small, lightweight items. Browne's reference to these materials' "electric" properties reflects the limited but growing curiosity about electrical phenomena at the time.

- **Work:** Browne, T. (1646). *Pseudodoxia Epidemica: or, Enquiries into Very many received Tenents, and commonly presumed Truths.* London: T. H. for Edward Dod, pp. 100-102.

- **Contributions:**

  - **Early Exploration of Attractive Forces:** Browne discussed the mysterious nature of attractive forces, referencing the behavior of materials like amber (which, when rubbed, attracts lightweight objects) and comparing it to the properties of the lodestone. His work reflects an early interest in the nature of electric attraction, which would later be formalized in the study of electrostatics.
  - **Speculative Theories on Magnetism:** Browne speculated on the relationship between magnetism and electricity, suggesting that the attractive properties of certain materials could be linked to their composition and inherent forces, though he did not have the experimental means to prove these ideas.
  - **Critical Examination of Electrical Phenomena:** Browne analyzed various "vulgar errors" related to magnetic and electric properties, aiming to demystify the misconceptions about the behavior of these forces. His work set the stage for later experimental studies in these fields.

Influence on Later Work: Sir Thomas Browne's early observations on attractive forces were precursors to the formal study of electrostatic phenomena in the 18th century. His work on dispelling misconceptions inspired later scientists to approach the study of natural forces with a critical and empirical mindset. Although his theories were largely speculative, they contributed to the broader discourse that eventually led to experimental verification by Stephen Gray and Benjamin Franklin.

> "For if rubbing of warm bodies emitteth a sweet fragrancy, or odor resembling that of the body, it is no marvel that it attracteth also the effluviums or steams of bodies adjacent."
>
> — Sir Thomas Browne, *Pseudodoxia Epidemica*, Book II, Chapter 3, p. 102.

2.4 1660s - Robert Boyle

Boyle's experimental approach and investigations of phenomena like electricity, magnetism, and air pressure.

- **Work:** *New Experiments Physico-Mechanical* (1660)

- **Contributions:**

  - **Experiments with Static Electricity:** In 1675, Boyle published Experiments and Notes About the Mechanical Origine or Production of Electricity. This work explored various materials' ability to generate static electricity when rubbed, building on earlier knowledge from the Greeks and contemporaries like William Gilbert. Boyle was one of the first to use the term "electric" to describe materials that exhibited attractive properties after being rubbed.

- **Magnetism and Magnetic Phenomena:** Boyle also conducted research on magnetism and magnetic materials. In his work New Experiments and Observations Touching Cold (1665), he describes experiments involving magnets and their interactions with other materials. His careful observations and experiments in this area contributed to the broader scientific understanding of magnetic forces, which would later be crucial for the study of electromagnetism.

- **Vacuum Experiments and the Role of the Medium:** Boyle is best known for his experiments with the air pump, which led to the formulation of Boyle's Law (relating the pressure and volume of a gas). In his vacuum experiments, Boyle also examined how air and the absence of a medium affected electrical and magnetic phenomena. For example, he observed that electrical attraction diminished in a vacuum, suggesting the importance of the medium in the transmission of electrical forces.

  These experiments raised fundamental questions about the nature of action at a distance and the role of the medium in transmitting forces, which were later central themes in the study of electromagnetism.

- **Boyle's Philosophical Contributions:** Boyle's emphasis on the mechanical philosophy and his advocacy for empirical experimentation influenced later scientists, including Isaac Newton. Newton's subsequent work on optics and gravity incorporated concepts that would eventually influence the unified theories of electromagnetism in the 19th century.

Influence on Later Work: Boyle's careful experimental methods, his exploration of static electricity, and his investigations into magnetic properties provided valuable data and theoretical insights that other scientists, such as Stephen Gray and Benjamin Franklin, would build upon in the 18th century. His work also indirectly influenced the 19th-century pioneers of electromagnetism, such as Michael Faraday and James Clerk Maxwell, who sought to understand the relationship between electricity and magnetism.

While Boyle did not create a unified theory of electromagnetism, his research helped establish key experimental techniques and foundational observations that contributed to the later development of the field.

> "It would be rash to conclude, that because a magnet does naturally attract iron, there must needs be in a magnet such a peculiar power, as is not explicable by any known mechanical principle."
>
> —Boyle, R.

2.5 1670s - Otto von Guericke

Otto von Guericke (1602–1686) was a German scientist, engineer, and politician best known for his pioneering work in vacuum physics and electrostatics. While Guericke did not directly work on electromagnetism as a unified field (which emerged only in the 19th century), his experiments in the 1670s contributed significantly to the understanding of electrostatic phenomena.

- **Work:** *Magdeburg Hemispheres* (1672)

- **Contributions:**

  - **Electrostatic Generator (1672):** Guericke's most notable contributions to the field of electricity was his invention of an electrostatic generator in 1672. This device was designed to generate static electricity using friction. Guericke constructed the generator by mounting a sulfur globe on an axis that could be rotated by hand. When the globe was rubbed by the hand or a cloth, it produced static electricity, causing it to attract lightweight objects like feathers and chaff.

    This sulfur globe experiment was one of the first deliberate attempts to produce and study electricity systematically. The generator was capable of producing visible sparks and demonstrated the attractive and repulsive properties of electrified objects, paving the way for more advanced electrostatic experiments.

  - **Glow Discharge in a Vacuum:** Guericke observed that when he rotated his sulfur globe in a darkened room, it emitted a faint glow, which he attributed to the presence of "electric effluvia." This was one of the earliest recorded observations of glow discharge, a phenomenon related to the behavior of electric charges in gases. Although he did not fully understand the underlying principles, this experiment contributed to the study of electrical discharges and would later be explored in more detail by other scientists like Michael Faraday.

  - **Electrostatic Attraction and Repulsion:** Guericke noted that the electrified sulfur globe could attract and hold small objects, similar to the way a magnet attracts iron filings. He also observed that, in some cases, the electrified objects repelled each other, depending on their charge distribution. This distinction between attraction and repulsion was a critical observation in understanding the nature of electrostatic forces and the concept of electrical polarity.

Influence on Later Work: Guericke's work was primarily focused on static electricity rather than magnetism, but his experiments significantly advanced the understanding of electrical phenomena. His sulfur globe became one of the first devices to generate a continuous and reproducible electric charge, allowing for systematic studies of electrostatic forces. This was a critical step in the broader field of electromagnetism because it provided a means to experiment with electrical forces and their effects. Gray's work on electrical conduction and insulation in the early 18th century built on Guericke's observations and experiments.Du Fay further explored the properties of electric charges, building on the phenomena that Guericke had observed.

He provided some of the earliest insights into electrical phenomena.

"The power of attracting and repelling in electrified bodies is greater when they are separated from each other by a vacuum."

—Guericke, O.

2.6 1680s - Isaac Newton

Isaac Newton (1642–1727) is best known for his work in mechanics, optics, and mathematics, but his contributions to electricity and magnetism were more indirect. During the 1680s, Newton's research did not focus directly on electromagnetism; however, his foundational ideas and explorations in natural philosophy paved the way for the later development of this field.

- **Work:** *Philosophie Naturalis Principia Mathematica* (1687)

- **Contributions:**

    - **Universal Gravitation:** While Newton's work primarily focused on gravity, his principles of universal gravitation and his mathematical approach to physical phenomena had a profound influence on scientific methodology. Newton's methods and concepts would later be applied to the study of electromagnetism, particularly in understanding the forces between charged particles. As forces decrease depending on distance. Differential equations would be required to understand electromagnetism.

    - **Optics and Light:** In the 1680s, Newton was deeply involved in studying light and optics, which culminated in the publication of his seminal work, Opticks (published in 1704, but reflecting research conducted in the late 17th century). Although Opticks primarily focuses on the nature of light, Newton also discusses how light might be related to other natural phenomena, including magnetism and electrical attraction.

        Newton proposed that light might consist of small particles and that these particles could interact with other forces. This speculation hinted at a possible connection between different types of forces—an idea that would be revisited when electromagnetism was unified in the 19th century by James Clerk Maxwell.

    - **Letter to Robert Boyle (1675):** While this letter predates the 1680s, it provides valuable insight into Newton's thinking about forces of attraction. In a letter to Robert Boyle in 1675, Newton expressed his belief that there might be an underlying unity between different forces of attraction, including gravitational, magnetic, and electric forces. He suggested that these forces might be governed by similar principles, even if the exact nature of their interactions was not yet understood.

    - **Hypotheses on Forces and Aether:** Newton also speculated about the existence of an "aether" as a medium for transmitting forces. In Opticks, he proposed that electrical and magnetic phenomena could be mediated through an aetherial medium. This hypothesis, although incorrect in the modern understanding, was an attempt to reconcile various forces of nature and hinted at a unified view of natural forces.

    - **Mathematical Framework:** Newton's Philosophiæ Naturalis Principia Mathematica (1687), commonly known as the Principia, introduced his laws of motion and the law of universal gravitation. While the Principia does not explicitly address electromagnetism, it established a rigorous mathematical framework for understanding forces and motion. This framework of forces acting at a distance influenced later physicists' approach to studying electric and magnetic forces.

In the Principia, Newton showed that gravitational force could be described mathematically and that it obeyed an inverse-square law. This was later found to be true for both electrical and magnetic forces as well. Coulomb's law for electrical forces (published by Charles-Augustin de Coulomb in 1785) follows the same mathematical form as Newton's law of gravitation, suggesting a possible analogy between these different forces.

Influence on Later Work: Although Newton did not conduct experiments directly on electromagnetism, his theories and ideas had a lasting impact.

Unified View of Forces: Newton's work on gravity and his mathematical description of forces acting at a distance set the stage for the study of electrical and magnetic forces as similar distant-acting forces. His suggestion that different types of forces might be connected influenced the subsequent search for a unified theory.

Newton's emphasis on empirical observation, mathematical rigor, and the search for universal laws of nature inspired later physicists like Michael Faraday, James Clerk Maxwell, and Albert Einstein. Faraday's field theory of electromagnetism and Maxwell's equations, which unified electricity and magnetism, can be seen as a continuation of Newton's vision for a unified understanding of natural forces.

"That bodies may be attracted to each other, it is requisite that they be not only in a state of rest but also that they be separated by a distance which is not infinite."

—Newton, I.

2.7 Summary of the 17th Century Contributions

Overall, the 17th century was a formative period that set the stage for the more detailed and systematic study of electromagnetism in the 18th and 19th centuries. The contributions of these early scientists and their methodologies played a crucial role in the evolution of electrical and magnetic theory.

- **Foundational Discoveries:** The 17th century saw the initial systematic studies of magnetism and static electricity, beginning with William Gilbert's *De Magnete*. This period marked the separation of magnetic and electrical phenomena and laid the groundwork for future discoveries.

- **Scientific Methodology:** Figures like Francis Bacon and Robert Boyle championed the scientific method and empirical research, which became central to scientific progress in electromagnetism.

- **Static Electricity:** Otto von Guericke's work with static electricity and early electrostatic generators provided practical insights and experimental evidence that contributed to the later development of electrical theory.

- **Influence of Classical Physics:** Isaac Newton's advancements in classical physics and his methodological approach influenced the broader scientific understanding that would encompass electromagnetism.

3 18th century

3.1 1729 - Stephen Gray:

Stephen Gray (1666–1736) was an English scientist. He made significant contributions to the field of electricity through his experiments with electrical conduction and insulation in the early 18th century. He discovered the difference between conductors and insulators, and he demonstrated the ability to transfer electricity over a distance using wires.

- **Work:** *Gray, S. (1731). "Experiments Concerning Electricity."* Philosophical Transactions of the Royal Society, 37, 18–44.

    Gray, S. (1732). "Further Experiments Concerning Electricity." Philosophical Transactions of the Royal Society, 37, 397–407.

- **Contributions:**

    - **Discovery of Electrical Conduction (1729):** Gray observed that electricity could travel through some materials (which he called conductors) while being stopped by others (which he called insulators). It demonstrated that electrical forces could be transmitted over a distance through a medium.

    - **Experiment with Suspended Cords:** His experiments involved suspending a cork ball by a silk thread and applying a charge to it using an electrified glass tube. He noticed that the charge could travel along the silk thread, causing the cork ball to attract other lightweight objects. He then extended the distance over which the charge could travel by using additional threads, sticks, and rods.

    - **Classification of Materials:** Gray classified materials into two categories: conductors, such as metal rods, moist threads, and wet wood, and insulators, such as silk, glass, and resin. His classification was one of the first systematic descriptions of the behavior of different materials when exposed to electric charges.

    - **The Concept of Insulators:** In addition to identifying conductors, Gray discovered that some materials could prevent the passage of electrical forces. He used silk threads and glass rods as insulating materials in his experiments to block the flow of electricity. This distinction between conductors and insulators remains fundamental in electrical theory and technology today.

    - **Transmission of Electric Charge Over Long Distances:** Gray demonstrated that electrical forces could be transmitted over relatively long distances—up to hundreds of feet—by using a series of suspended rods and threads. This finding was a crucial step in understanding how electric forces propagate, foreshadowing the concept of electric circuits and the flow of electricity.

- **"The Flying Boy" Experiment (1729)**: One of Gray's famous experiments was the "Flying Boy" experiment, where he suspended a young boy horizontally using silk ropes. The boy was then electrified, and objects like small pieces of paper or feathers were attracted to his body. This visually demonstrated the principle that the charge could be transmitted through the suspended silk ropes without direct contact with the electrifying device. The experiment captured public attention and served as a striking demonstration of electrical conduction and the role of insulating materials.

Influence on Later Work: Gray's experiments had a profound influence on later scientists, such as Charles Dufay and Benjamin Franklin. Dufay's work on distinguishing between charges, while Franklin further explored the nature of electric charge and its transmission.

His discoveries about electrical conduction and the behavior of electric forces were essential for the later unification of electricity and magnetism. His experiments showed that electrical forces could act over distances and through various materials, providing a foundation for understanding how electric and magnetic fields propagate.

Michael Faraday's work on electromagnetic induction and James Clerk Maxwell's formulation of Maxwell's equations, which describe the relationship between electric and magnetic fields, were built on the fundamental principles that Gray helped to establish.

"By the help of the Electric Glass, and some other materials, I have made a good many Experiments, and shall now lay them before you."

—Gray, S.

3.2 1733 - Charles François de Cisternay du Fay

Dufay (1698–1739) was a French chemist and physicist. He is most famous for his discovery of the existence of two distinct types of electric charges, which he called "vitreous" and "resinous" electricity. This concept laid the foundation for what would later be understood as positive and negative charges.

- **Work:** Du Fay, C. F. (1733). "Sur l'électricité produite par frottement." Mémoires de l'Académie Royale des Sciences, 23, 23–35.

  Du Fay, C. F. (1734). "Sur les différentes sortes d'électricité." Mémoires de l'Académie Royale des Sciences, 24, 439–451.

- **Contributions**:

  - **Discovery of Two Types of Electric Charge (1733)**: One of Du Fay's most important contributions to electrical science was his discovery that there are two kinds of electric charges. In a series of experiments, he observed that certain materials, such as glass, could be electrified in a way that attracted other materials, like paper or feathers. However, he noticed that materials electrified by different substances sometimes repelled each other rather than attracting.

- **Vitreous and Resinous Electricity:** Du Fay called the type of electricity generated by rubbing glass vitreous electricity and the type generated by rubbing resin resinous electricity. He demonstrated that materials charged with vitreous electricity would repel each other but attract materials charged with resinous electricity, and vice versa. This was the first identification of two distinct types of electric charge, a concept that was crucial to later work in electromagnetism.

- **Explanation of Attraction and Repulsion:** Du Fay explained that like charges repel each other, while unlike charges attract. This principle of attraction and repulsion became a cornerstone of electrical theory and was later formalized in Coulomb's law, which describes the force between two electric charges.

- **The Principle of Electrical Neutrality:** Du Fay's experiments led to the principle that materials in their natural state are electrically neutral, having equal quantities of both types of electricity. He showed that when a material becomes electrified, it gains one type of electricity and loses the other, resulting in an imbalance that causes attraction or repulsion with other electrified bodies. This idea hinted at the concept of charge conservation and the nature of electric polarization.

- **Conduction and Insulation:** Building on the earlier work of Stephen Gray, Du Fay studied how electricity could be conducted through different materials. He further confirmed Gray's findings by classifying materials into conductors (substances that allowed electricity to pass through them, like metals) and insulators (substances that did not allow electricity to pass through, like glass and silk).

- **Extension of Electrical Conduction:** Du Fay showed that the same principles of conduction and insulation applied to both types of electricity (vitreous and resinous). He demonstrated that a body could be charged with one type of electricity through contact with a charged conductor, and that this charge could be transmitted or blocked depending on the materials used.

- **Du Fay's Experiments with Electroscopes:** Du Fay used electroscopes—devices that detect electric charge—to observe the behavior of electrified bodies. His use of electroscopes helped him differentiate between the two types of electricity and study their interactions in a more systematic manner.

- **Experiments on the Influence of Environment on Electrification:** Du Fay conducted experiments to understand how environmental factors, like humidity, affected electrification. He found that humid air reduced the ability of materials to become electrified, which he attributed to moisture acting as a conductor and dissipating electric charges. This insight contributed to the understanding of how external conditions influence electrical behavior, which is still relevant in modern studies of dielectric materials and insulation.

Influence on Later Work: Although Du Fay's work was focused on electricity, his discoveries were foundational for the later unification of electricity and magnetism. His identification of two types of charges provided a framework that allowed scientists like Benjamin Franklin to build a more comprehensive theory of electricity. Du Fay's research on attraction and repulsion was later formalized by Coulomb. Michael Faraday's field theory of electricity and magnetism also relied on understanding the interactions between different types of charges. The ideas of electric charge, conduction, and polarization that Du Fay explored were essential to the later formulation of Maxwell's equations in the 19th century.

"There are two kinds of electricities, which may be called vitreous and resinous, and they repel each other."

—Du Fay, C. F.

3.3 1745-1746 Ewald Georg von Kleist and Pieter van Musschenbroek

Ewald Georg von Kleist (1700–1748) and Pieter van Musschenbroek (1692–1761) independently invented the Leyden jar, one of the earliest devices for storing electric charge, in the mid-18th century. Their work laid a crucial foundation for the study of electrostatics.

- **Work:** Development of the Leyden jar

    Kleist, E. G. von. (1745). Description of a New Method to Store Electricity. Personal correspondence and contemporary reports (unpublished work at the time but documented in historical accounts).

    Musschenbroek, P. van. (1746). "Epistola de Magno Experimento Electricitatis." Mémoires de l'Académie Royale des Sciences, Volume 20, 1–8.

- **Contributions:**

    – **Leyden Jar:** Independently, both von Kleist (in Germany) and van Musschenbroek (in the Netherlands) developed the Leyden jar, an early form of a capacitor that could store electrical charge. The Leyden jar consisted of a glass jar lined with metal foil inside and outside, with an insulating layer of glass in between.

    – **Observation:** Kleist found that holding the jar in his hand, he received a powerful shock when the wire was touched, thus demonstrating that the device stored electrical energy. In one of Musschenbroek's famous experiments, he held the jar and noted that when he touched the wire, he experienced a severe shock, indicating the successful storage and discharge of electricity.

    – **Capacitance and Storage:** The Leyden jar demonstrated the ability to store a static electrical charge and was used in experiments to study the discharge of electrical energy. This invention was a key development in the study of electricity and laid the groundwork for later electrical components.

Influence on Later Work: Although the Leyden jar itself is an electrostatic device, it set the stage for subsequent exploration of electricity and magnetism. The ability to store electrical energy and control its discharge enabled scientists like Benjamin Franklin to conduct systematic experiments. These experiments provided insights into the behavior of electricity and led to the development of key concepts like electric potential, capacitance, and the flow of electric charge.

In the 19th century, the principles of energy storage and electrical discharge discovered using devices like the Leyden jar played a role in the broader understanding of electromagnetic theory, especially in relation to oscillating circuits and electromagnetic waves.

"I discovered that when I had taken a glass jar and filled it with water, and electrified it, it could give a shock."

—Kleist, E. G.

"This jar may be charged with an electrical force, and from it, one can draw sparks, thereby demonstrating the power of electricity."

—Musschenbroek, P. v.

3.4 1752 - Benjamin Franklin

Benjamin Franklin (1706–1790) was born in America, he was a pioneering figure in the study of electricity. His experiments and theoretical contributions during the mid-18th century had a profound impact. Franklin's work focused primarily on electricity.

- **Work:** *Franklin, B. (1751). "Experiments and Observations on Electricity Made at Philadelphia in America."* London: E. Cave. This work consists of a series of letters and experiments Franklin conducted on electricity, including his findings on positive and negative charges, conservation of charge, and lightning as electricity.

  Franklin, B. (1752). "Letter to Peter Collinson." Published in the Philosophical Transactions of the Royal Society, detailing the famous kite experiment and his observations on the nature of lightning.

- **Contributions:**

  - **Kite Experiment:** Franklin's famous kite experiment in 1752 provided evidence that lightning is a form of electricity. He showed that electrical charge could be conducted from the atmosphere to a grounded conductor. He flew a kite during a thunderstorm, allowing a metal key attached to the kite string to become electrified by the storm's static electricity. When Franklin brought his knuckle near the key, he observed a spark, which proved that the key had become charged with the electrical energy of the storm. This confirmed his hypothesis that lightning is an electrical discharge.

  - **Electricity Theory:** Franklin introduced the concepts of positive and negative electrical charges and formulated the principle of conservation of charge. His work on the nature of electrical charge and electric potential contributed significantly to the theoretical understanding of electricity.

- **The Concept of Positive and Negative Charges:** Franklin proposed the concept of positive and negative charges, which was a significant refinement of Charles Dufay's theory of two types of electricity (vitreous and resinous). Franklin introduced the notion of electrical fluid and hypothesized that electricity was a single fluid that could be transferred between objects, leading to an excess (positive) or a deficit (negative) of electrical fluid. He argued that when an object loses electrical fluid, it becomes negatively charged, and when it gains electrical fluid, it becomes positively charged. This was a major advancement in understanding electric charge as a transferable entity.

- **The Principle of Conservation of Charge:** Franklin's experiments led him to propose the principle of conservation of charge, stating that the total amount of electric charge in an isolated system is constant. This means that when one object is charged, it does not create new charge but simply redistributes existing charge. This idea was critical in forming the foundation of modern electrical theory.

- **The One-Fluid Theory of Electricity:** Franklin's one-fluid theory of electricity replaced the previous two-fluid theories of his contemporaries. He proposed that electricity consisted of a single fluid, and the interaction between positive and negative charges could explain all electrical phenomena. This theory helped systematize the study of electricity and contributed to the development of a more unified understanding of electric and magnetic forces.

- **Identification of Electric Conduction and Polarization:** Franklin also investigated the behavior of conductors and insulators, defining conductors as materials that allow the free flow of electrical fluid and insulators as those that do not. He explored the polarization of charges on the surface of conductors, leading to a better understanding of electrostatic induction.

Influence on Later Work: His principles of electric charge, conservation of charge, and conduction were essential for later developments. The Kite experiment influenced the invention of lightning rod, a device that protects buildings from lightning strikes by providing a path for electrical discharge to reach the ground safely.

Franklin's work on electric force and his observations of attraction and repulsion influenced Charles-Augustin de Coulomb's formulation of Coulomb's law Franklin's concepts of electric charge distribution and electric fields were foundational for Michael Faraday's field theory of electromagnetism and James Clerk Maxwell's unification of electric and magnetic fields in his set of equations.

"It is the opinion of many that the electric fluid is the cause of the attraction of bodies."

—Franklin, B.

3.5   1770s - Joseph Priestley

Joseph Priestley (1733–1804), best known as a chemist and the discoverer of oxygen, also made important contributions to the study of electricity. His work in the 1770s focused on the nature of electrical forces and the interaction between electrically charged bodies influenced the development of electromagnetic theories.

- **Work:** Experiments on gases and electricity

    Priestley, J. (1767). "The History and Present State of Electricity, with Original Experiments." London: J. Dodsley.

    This is Priestley's major publication on electricity, where he presents his ideas about the inverse square law, electrical fluids, and his experiments with conductors. It also contains a historical survey of the development of electrical science up to his time.

    Priestley, J. (1775). "Observations on Different Kinds of Air and the Inverse Square Law."

    Published in various letters and scientific papers, this work outlines his theoretical insights on electrical force and distance.

- **Contributions:**

    - **Charged hollow sphere:** He showed that an electrically charged hollow sphere, when brought near other objects, does not transmit any electrical force inside the hollow cavity, a concept that is critical in understanding the behavior of electric fields and the principle of electrostatic shielding.

    - **Observation:** Priestley reasoned that if an electrical body can exert no force inside a hollow conductor (as found in his experiments), then the electrical force outside must decrease as the square of the distance from the charge. Concept of Electrical Shielding: Priestley's experiments with hollow spherical conductors led to his observation of what is now known as electrical shielding. He noticed that when an electrically charged object is placed inside a hollow conductor, the electrical forces are only present on the exterior surface of the conductor, while the interior remains unaffected.

    - **Relationship Between Electricity and Matter:** Priestley speculated about the relationship between electricity and the atomic structure of matter. He suggested that electrical forces might be linked to the fundamental structure of materials, anticipating some of the ideas that would later lead to the understanding of electric fields and their effects on matter.

    - **Electrolysis of Water (1774):** Priestley conducted experiments that involved the electrolysis of water, breaking it down into hydrogen and oxygen gases.

    - **Support for Franklin's One-Fluid Theory:** Priestley supported Benjamin Franklin's one-fluid theory of electricity and further elaborated on the nature of electrical charges. He used Franklin's theory as a basis to explain various electrical phenomena, including induction and the attraction and repulsion of charges.

Influence on Later Work: In his seminal work, The History and Present State of Electricity, the inverse-square law of electrical force that he proposed is a direct analogue to

Newton's law of gravitation and was later experimentally verified by Coulomb. This law is a fundamental principle in the study of electric fields and forces.

Priestley's observations on electrical shielding and the distribution of charges in conductors were important precursors to the mathematical treatment of electrostatics and the field theory developed.

"Electricity is a subtle fluid, which we have not yet learned to define."

—Priestley, J.

3.6 1785 - Charles-Augustin de Coulomb

Charles-Augustin de Coulomb (1736–1806) was a French physicist and engineer whose work in the late 18th century laid the foundation for the quantitative study of electrostatics.

- **Work:** Coulomb's Law

  Coulomb, C. A. (1785). "Premier Mémoire sur l'Électricité et le Magnétisme: De la quantité d'électricité produite dans un conducteur de forme donnée, soit régulièrement soit irrégulièrement arrondi." Histoire de l'Académie Royale des Sciences, 569–577.

  Coulomb, C. A. (1785). "Second Mémoire sur l'Électricité et le Magnétisme: Application des lois de l'électricité au magnétisme." Histoire de l'Académie Royale des Sciences, 578–611.

  Coulomb, C. A. (1785). "Troisième Mémoire sur l'Électricité et le Magnétisme: Force de torsion et force répulsive de l'électricité." Histoire de l'Académie Royale des Sciences, 612–638.

- **Contributions:**

  - **Coulomb's Law (1785):** Coulomb's most famous contribution is the formulation of Coulomb's Law, which describes the force between two point charges. He determined that the electric force between two charges is directly proportional to the product of the magnitudes of the charges and inversely proportional to the square of the distance between them. Coulomb's Law is analogous to Newton's law of gravitation but applies to electric charges instead of masses. This law provided a precise mathematical relationship for electric forces, making it possible to calculate the behavior of charges in an electric field.

  - **Experimental Verification:** Coulomb used a torsion balance is a sensitive instrument, which he invented to measure the forces between charged objects. By observing the angle of deflection in the torsion wire, he could quantify the force between the charges. This innovative technique allowed him to confirm the inverse-square nature of the force between charges.

  - **Quantitative Study of Magnetic Forces:** Coulomb also extended his studies to magnetic forces. He demonstrated that the force between magnetic poles follows a similar inverse-square law, analogous to the electric force. His experiments with magnets and magnetic materials provided a quantitative framework for the study of magnetism, setting the stage for the later unification of electricity and magnetism.

- **Magnetic Law:** The force between two magnetic poles is directly proportional to the product of their pole strengths and inversely proportional to the square of the distance between them, similar to Coulomb's Law for electric charges.

- **The Concept of Electric Charge and Field:** Coulomb's work helped define the concept of electric charge and electric field, providing a clear distinction between the quantity of charge and the electric force it exerts. His studies led to the notion of the electric field as a space around a charged object where electric forces are exerted.

- **Torsion Balance and Experimental Methods:** His method of quantifying the forces between charges and magnetic poles allowed for the establishment of fundamental constants and the validation of theoretical laws with experimental data.

- **Contributions to Electrostatics and Magnetostatics:** In addition to his laws of force, Coulomb studied the behavior of electric charges on conductors and the distribution of charge on surfaces. He explored the effects of shape, size, and material properties on the behavior of charges. In magnetostatics, he studied the interaction of magnetic dipoles and the behavior of magnetic materials, contributing to the understanding of magnetic fields and their sources.

Influence on Later Work: Coulomb's Law is a cornerstone of electromagnetism and is used in Maxwell's equations, which describe the behavior of electric and magnetic fields. His work on magnetic forces also provided a basis for the later development of magnetic field theory. Coulomb's precise definition of electric and magnetic forces influenced others to understand the dynamic relationship between electricity and magnetism.

"The forces of attraction and repulsion between electric charges are inversely proportional to the square of the distance between them."

—Coulomb, C.-A. de

3.7 1791 - Luigi Galvani

Luigi Galvani (1737–1798) was an Italian physician and physiologist who made significant contributions to the field of bioelectromagnetics through his discovery of what he called "animal electricity." His work laid the foundation for the study of electrophysiology and led to the later development of galvanism, influencing both scientific thought and experimental approaches in the field of electromagnetism.

- **Work:** Galvani, L. (1791). *De Viribus Electricitatis in Motu Musculari Commentarius* (Commentary on the Effect of Electricity on Muscular Motion). Bologna: Ex Typographia Instituti Scientiarum.

    Galvani, L. (1794). "Essay on the Effects of Electricity on Muscular Motion." *Philosophical Transactions of the Royal Society*, 84, 394–407.

- **Contributions:**

  - **Discovery of Animal Electricity (1780s):** Galvani discovered that the muscles of dead frogs' legs twitched when struck by a spark of static electricity. This phenomenon, which he called "animal electricity," suggested that electricity was involved in nerve signals and muscular movement.

  - **The Frog Leg Experiment (1791):** In his famous experiment, Galvani hung frog legs from an iron railing by brass hooks. He observed that the frog legs contracted whenever they touched the iron, even without an external source of electricity. This led him to hypothesize that the tissues themselves contained an intrinsic electrical energy, which was released upon contact between the different metals.

  - **Theory of Intrinsic Electrical Energy in Tissues:** Galvani theorized that animal tissues contained an innate form of electrical energy, which he termed "animal electricity." He believed that the brain sent electrical signals through nerves to activate muscles, suggesting that electricity was integral to physiological processes.

  - **Introduction of Galvanism:** The concept of "galvanism," named in his honor, emerged from Galvani's experiments. It refers to the direct stimulation of muscular and nerve tissue using electrical currents. His research opened the door to new fields of study, including bioelectromagnetics and electrophysiology.

  - **Controversy with Alessandro Volta:** Galvani's conclusions led to a scientific debate with Alessandro Volta, who argued that the observed muscle contractions were caused by the interaction of the metals rather than intrinsic electricity in the tissues. This debate eventually led Volta to invent the voltaic pile, but it also underscored the importance of Galvani's initial findings in demonstrating the relationship between electricity and biological systems.

Influence on Later Work: Galvani's discovery of animal electricity had a profound impact on the scientific community, influencing not only the study of electricity but also the emerging field of neuroscience. His experiments provided early evidence that electrical signals are involved in neural and muscular activity, a concept that was further developed by later scientists such as Emil du Bois-Reymond and Hermann von Helmholtz in their studies of nerve conduction.

Galvani's work also inspired Alessandro Volta to investigate the electrical properties of metals, leading to the invention of the voltaic pile, the first true battery. While Volta's theory of contact electricity opposed Galvani's animal electricity hypothesis, the scientific debate between them accelerated the understanding of electricity and magnetism.

Mary Shelley's novel *Frankenstein* (1818) was inspired by Galvani's work, reflecting the cultural and scientific fascination with the possibility of reanimating life through electrical stimulation. His research also laid the groundwork for modern medical applications, such as electrical stimulation therapy and the development of devices like pacemakers and defibrillators.

"The electric force seems to act in the same way as the nerve, which can be excited by a stimulus."

—Galvani, L.

3.8 Summary of the 18th Century Contributions

- **Discovery and Classification:** The 18th century saw the discovery of key concepts such as conductors and insulators, the development of the Leyden jar, and the demonstration of electrical charge through experiments like Franklin's kite experiment.

- **Mathematical Formulation:** Coulomb's Law provided a mathematical framework for understanding electrostatic forces, which was crucial for the later development of electromagnetic theory.

- **Foundational Work:** The contributions of scientists like Stephen Gray, Benjamin Franklin, and Charles-Augustin de Coulomb laid the groundwork for the formal study of electromagnetism and set the stage for the more integrated theories of the 19th century.

Overall, the 18th century marked a period of significant progress in the study of electricity and magnetism. The experiments, discoveries, and theoretical developments during this time laid the critical groundwork for the unification of electrical and magnetic phenomena that would be achieved in the 19th century.

4 19th Century: The Unification of Electricity and Magnetism

4.1 1800 - Alessandro Volta

Alessandro Volta (1745–1827) was an Italian physicist and chemist. Volta invented the first chemical battery, which provided a steady electric current. His research on electrical potential and his invention of the electric battery was crucial for studying modern electrochemistry and the understanding of electrical circuits. His studies helped bridge the gap between chemistry and electricity, demonstrating that chemical reactions can produce electrical energy.

- **Work:** Volta, A. (1800). "On the Electricity Excited by the Mere Contact of Conducting Substances of Different Kinds." *Philosophical Transactions of the Royal Society*, 90, 403–431.

  Volta, A. (1801). "On the Electricity Excited by the Contact of Metals with a Wet Surface." *Philosophical Transactions of the Royal Society*, 91, 427–429.

  These seminal papers document Volta's experimental results and his groundbreaking inventions.

- **Contributions:**

- **Invention of the Voltaic Pile (1800):** Volta's most significant contribution was the invention of the voltaic pile, the first true battery. The voltaic pile consisted of alternating discs of zinc and copper, separated by cardboard soaked in saltwater. This arrangement generated a continuous and stable electric current, a groundbreaking discovery that contradicted the prevailing belief that electricity could only be produced by living organisms (animal electricity).

- **Establishing the Concept of Electrical Potential:** Volta introduced the concept of electrical potential (voltage) and potential difference, which are key to understanding electric circuits and energy transfer. He showed that electricity could be generated simply by placing two different metals in contact, thus producing a potential difference.

- **The Law of Electromotive Force:** Volta's work on the voltaic pile led to the formulation of the law of electromotive force (EMF), which describes how chemical reactions in the battery produce a flow of electric charge. This concept is fundamental in electrochemistry and forms the basis for understanding how batteries and cells function.

- **Challenge to Animal Electricity Theory:** Volta's experiments challenged the then-dominant theory of animal electricity, proposed by Luigi Galvani, which suggested that electricity originated from animal tissues. Volta demonstrated that the electrical current could be generated without any biological material, using only metal and a conductive solution. This finding was pivotal in separating the study of electricity from biology and establishing electrochemistry as a distinct field.

- **Discovery of Contact Electricity:** Volta's exploration of the contact between different metals led to the discovery of contact electricity, or the phenomenon where a potential difference is produced at the junction of two dissimilar metals. This principle was later expanded upon in studies of thermoelectricity and semiconductor junctions.

- **Early Understanding of Electric Currents:** By constructing a device that could produce a steady electric current, Volta provided the first practical source of electricity. This development enabled future research in electromagnetism, electrochemistry, and the design of electrical circuits.

Influence on Later Work: Volta's invention of the voltaic pile and his studies of electrical potential had a profound influence on subsequent research in electromagnetism. His work provided the first reliable source of electrical energy, making it possible to perform experiments that required a continuous current, which was not possible before.

Michael Faraday's experiments on electromagnetism, including the discovery of electromagnetic induction, were directly facilitated by the availability of a steady current source provided by the voltaic pile.

> "An electric current, if properly directed, can produce various effects, including the ability to generate movement."
>
> —Volta, A.

4.2 1820 - Hans Christian Ørsted

Hans Christian Ørsted (1777–1851) was a Danish physicist and chemist, best known for his discovery of the relationship between electricity and magnetism. His seminal experiment in 1820 demonstrated that an electric current could create a magnetic field, providing the first empirical evidence of the link between electrical and magnetic phenomena. This discovery marked the beginning of the field of electromagnetism and influenced numerous subsequent developments in both theoretical and experimental physics.

- **Work:** Ørsted, H. C. (1820). *"Experimenta Circa Effectum Conflictus Electrici in Acum Magneticam"* (Experiments on the Effect of a Current of Electricity on the Magnetic Needle). *Annals of Philosophy*, 16, 273–276.

  Ørsted, H. C. (1820). *"Experiments on the Effect of a Current of Electricity on the Magnetic Needle."* *Edinburgh Philosophical Journal*, 2, 381–385.

- **Contributions:**

  - **Discovery of Electromagnetism (1820):** Ørsted's most important contribution came from his experiment demonstrating that an electric current produces a magnetic field. During a lecture, he noticed that a compass needle deflected from magnetic north when placed near a conducting wire carrying an electric current. This phenomenon provided direct evidence of a connection between electricity and magnetism.

  - **Formulation of the Electromagnetic Effect:** Ørsted observed that the magnetic field produced by the current was circular, surrounding the conductor in concentric circles. This led to the realization that the magnetic field was not aligned along the wire, but rather formed a continuous loop around it. His observations paved the way for the mathematical description of the magnetic field generated by electric currents.

  - **Establishment of the Concept of Electromagnetic Fields:** By showing that a magnetic field could be generated by an electric current, Ørsted introduced the concept of the electromagnetic field—a unified region where electric and magnetic forces are interrelated. This concept is a cornerstone in the study of classical electromagnetism.

  - **Influence on the Development of Electromagnetic Theory:** Ørsted's discovery directly inspired other scientists, such as André-Marie Ampère and Michael Faraday, to further investigate the nature of the relationship between electricity and magnetism. Ampère's work, in particular, led to the formulation of Ampère's Law, which describes the magnetic field produced by a current-carrying wire in quantitative terms.

  - **The Ørsted Experiment:** The setup involved a conducting wire, a source of electric current, and a magnetic compass needle. When the current flowed through the wire, the needle deflected, indicating the presence of a magnetic field. This experiment demonstrated the first practical example of electromagnetism and provided a foundation for understanding how electric currents generate magnetic fields.

Influence on Later Work: Ørsted's discovery of electromagnetism was a pivotal moment in the history of physics, leading to the unification of electricity and magnetism into a single theory. His work directly influenced Ampère's mathematical description of electromagnetism. Michael Faraday's research on electromagnetic induction was also inspired by Ørsted's findings.

Maxwell incorporated Ørsted's discovery into his equations, which describe the behavior of electric and magnetic fields. Ørsted's influence extended as the concepts of electric and magnetic fields became integral to technologies such as electric motors, generators, and transformers.

"When an electric current passes through a wire, it produces a magnetic effect around it."

—Ørsted, H. C.

4.3 1820 - André-Marie Ampère

André-Marie Ampère (1775–1836) was a French physicist and mathematician. He is often regarded as one of the founders of electrodynamics. Ampère's work established the theoretical and mathematical foundation of electromagnetism, formalizing the relationship between electric currents and magnetic fields. His formulation of Ampère's Law and the development of the concept of the electromagnetic field were instrumental in advancing the understanding of electromagnetic phenomena.

- **Work:** Ampère, A.-M. (1820). *"Mémoire sur l'action mutuelle de deux courants électriques"* (Memoir on the Mutual Action of Two Electric Currents). *Annales de Chimie et de Physique*, 15, 59–76.

  Ampère, A.-M. (1820). *"Théorie des phénomènes électrodynamiques, uniquement déduite de l'expérience"* (Theory of Electrodynamic Phenomena, Deduced Solely from Experiment). *Annales de Chimie et de Physique*, 20, 496–524.

- **Contributions:**

    - **Discovery of the Electrodynamic Force Law (1820):** Ampère discovered that two parallel conductors carrying electric currents exert forces on each other. He observed that currents flowing in the same direction attract, while those flowing in opposite directions repel each other. This relationship, known as Ampère's Force Law, provided a quantitative description of the interaction between electric currents.

    - **Formulation of Ampère's Law:** Ampère's Law relates the magnetic field in space to the electric current that produces it. It states that the integrated magnetic field around a closed loop is proportional to the total current passing through the loop. This mathematical formulation became one of Maxwell's equations, which describe the behavior of electric and magnetic fields.

- **Introduction of the Concept of Electrodynamic Potentials:** Ampère introduced the concept of electrodynamic potentials, which allowed him to express the forces between current-carrying conductors in a more general form. His development of these potentials laid the groundwork for future theoretical advancements in electromagnetism.

- **Establishment of Electrodynamics as a Field of Study:** Ampère's systematic study of the interactions between electric currents and magnetic fields established electrodynamics as a distinct branch of physics. His work defined the rules for calculating the forces and torques between current-carrying wires, leading to the development of numerous technologies based on electromagnetic principles.

- **The Ampère Experiment (1820):** Ampère designed a series of experiments to measure the force between two parallel current-carrying wires. He used a torsion balance to quantify the attractive and repulsive forces, providing experimental validation for his theoretical formulations. This experiment marked the first quantitative study of the forces between electric currents.

Influence on Later Work: His discovery of the force law between currents provided the theoretical framework that guided subsequent research. Faraday, inspired by Ampère's work, conducted experiments that led to the discovery of electromagnetic induction, a key principle in electromagnetism. Ampère's Law was later incorporated into Maxwell's equations.

Ampère's theoretical advancements also played a crucial role in the development of electric motor technology and the study of magnetic materials. His work on the interactions between currents and magnetic fields set the stage for the modern understanding of electromagnetic fields and waves.

"The current which is in the same direction in two parallel conductors attracts one another, and if it is in opposite directions, it repels one another."

—Ampère, A.-M.

4.4 1831 - Michael Faraday

Michael Faraday (1791–1867) was an English scientist whose groundbreaking research in electromagnetism and electrochemistry revolutionized the understanding of electrical and magnetic phenomena. Faraday's experiments in electromagnetic induction and the discovery of the laws of electrolysis laid the foundation for many modern technologies, including electric generators, transformers, and inductors. His work demonstrated the principles of electromagnetic fields, leading to the unification of electricity and magnetism into a single coherent theory.

- **Work:** Faraday, M. (1831). *"On the Induction of Electric Currents." Philosophical Transactions of the Royal Society,* 121, 125–162.

  Faraday, M. (1832). *"Experimental Researches in Electricity." Philosophical Transactions of the Royal Society,* 122, 163–186.

- **Contributions:**

  - **Discovery of Electromagnetic Induction (1831):** Faraday discovered that a changing magnetic field could induce an electric current in a conductor. This phenomenon, known as electromagnetic induction, is the principle behind electric generators and transformers. He demonstrated that moving a magnet through a coil of wire produced an electric current in the wire, thus establishing the concept of the electromagnetic field.

  - **Faraday's Law of Induction:** Faraday formulated the law of electromagnetic induction, which states that the magnitude of the induced electromotive force (EMF) in a circuit is proportional to the rate of change of the magnetic flux through the circuit. This law quantitatively describes how electric currents are generated by changing magnetic fields and is one of the fundamental principles of electromagnetism.

  - **Introduction of the Concept of Field Lines:** Faraday introduced the concept of electric and magnetic field lines to represent the influence of electrical and magnetic forces in space. He visualized the field as a series of lines emanating from electric charges or magnets, providing a qualitative method to understand the direction and strength of electric and magnetic fields.

  - **Discovery of Diamagnetism and Paramagnetism (1845):** Faraday discovered that materials could be classified as diamagnetic or paramagnetic based on their response to a magnetic field. Diamagnetic materials are repelled by a magnetic field, while paramagnetic materials are attracted. This classification contributed to the understanding of how materials interact with magnetic fields.

  - **Faraday's Cage Experiment (1836):** Faraday demonstrated that an enclosed conducting shell can shield its interior from external static electric fields. This experiment led to the development of the concept of the "Faraday cage," which is used in various applications to block external electromagnetic interference.

  - **Discovery of the Relationship Between Light and Electromagnetism (1845):** Faraday discovered that a magnetic field could influence the polarization of light, an effect known as the Faraday Effect. This experiment demonstrated that light itself is an electromagnetic phenomenon, laying the groundwork for James Clerk Maxwell's theory of electromagnetism.

Influence on Later Work: Faraday's experiments and conceptualization of electromagnetic fields helped the development of classical electrodynamics. His discovery of electromagnetic induction directly influenced the work of Maxwell, who used Faraday's findings to develop Maxwell's Equations. These equations mathematically describe the behavior of electric and magnetic fields and their interdependence, forming the basis of classical electromagnetism.

Faraday's visualization of electric and magnetic fields introduced the concept of the field, which became central to the modern understanding of physical interactions. His work on the Faraday Effect provided early evidence of the electromagnetic nature of light, which

Maxwell later confirmed in his unified theory of electromagnetism. Faraday's influence extended beyond physics, as his discoveries laid the groundwork for numerous technological advancements, including the development of electrical power generation and the study of electromagnetic waves.

> "When a conductor is moved through a magnetic field, an electric current is induced in the conductor."
>
> —Faraday, M.

4.5 1864 - James Clerk Maxwell

James Clerk Maxwell (1831–1879) was a Scottish physicist who is renowned for his formulation of the classical theory of electromagnetic radiation. Maxwell's four equations, known collectively as Maxwell's Equations, unified the study of electricity, magnetism, and optics. His work established the theoretical framework for understanding how electric and magnetic fields interact and propagate as electromagnetic waves. Maxwell's contributions to electromagnetism fundamentally changed the field and laid the foundation for modern physics.

- **Work:** Maxwell, J. C. (1864). "*A Dynamical Theory of the Electromagnetic Field.*" *Philosophical Transactions of the Royal Society of London*, 155, 459–512.

  Maxwell, J. C. (1873). *A Treatise on Electricity and Magnetism*. Oxford: Clarendon Press.

- **Contributions:**

  - **Formulation of Maxwell's Equations (1861–1862):** Maxwell unified previous observations and empirical laws of electricity and magnetism, including those of Coulomb, Ampère, and Faraday, into a set of four partial differential equations. These equations mathematically describe the behavior of electric and magnetic fields and their interdependence. Maxwell's Equations established that electric and magnetic fields propagate as waves at the speed of light.

  - **Prediction of Electromagnetic Waves:** Maxwell's Equations predicted the existence of electromagnetic waves, showing that oscillating electric and magnetic fields can propagate through space. He calculated the speed of these waves and found it to be equal to the speed of light, leading to the conclusion that light itself is an electromagnetic wave.

  - **Theory of Electromagnetic Radiation:** In his 1864 paper, Maxwell proposed that electromagnetic fields could propagate through space without the need for a medium. This contradicted the prevailing theory of the "luminiferous aether" and suggested that light, radio waves, and X-rays were all manifestations of the same electromagnetic phenomenon.

  - **Unification of Electricity, Magnetism, and Optics:** Maxwell's work showed that electric fields, magnetic fields, and light are all interrelated aspects of the same underlying physical reality. His equations demonstrated that a changing magnetic field produces an electric field and vice versa, leading to a comprehensive theory that unified the three fields.

- **Displacement Current and Modification of Ampère's Law:** Maxwell introduced the concept of the "displacement current" to account for the changing electric field in a capacitor. This addition to Ampère's Law allowed for the prediction of electromagnetic wave propagation and was essential for the consistency of the equations.

- **Maxwell's Treatise on Electricity and Magnetism (1873):** Maxwell's treatise is considered one of the most important texts in the history of physics. It systematically presented his theories on electromagnetism and introduced new concepts such as the electromagnetic field and the potential function. The book influenced generations of physicists and became a cornerstone for the development of classical electrodynamics.

Influence on Later Work: Maxwell's equations provided the theoretical foundation for much of modern physics, influencing the work of scientists such as Heinrich Hertz, who experimentally confirmed the existence of electromagnetic waves. Maxwell's theory laid the groundwork for Albert Einstein's theory of special relativity, which redefined concepts of space and time by integrating Maxwell's equations with the principles of mechanics.

The field of electrodynamics, as defined by Maxwell, became the basis for numerous technological advancements, including the development of radio, television, and wireless communication. Maxwell's insights into the nature of light and electromagnetic waves also paved the way for the discovery of quantum mechanics and the study of electromagnetic interactions at the quantum level.

> "The phenomena of electrostatics and electrodynamics are not only intimately connected, but are actually parts of one and the same science."
>
> —Maxwell, J. C.

5 Late 19th Century to Early 20th Century: Technological Applications

5.1 1887 - Heinrich Hertz

Heinrich Hertz (1857–1894) was a German physicist who made significant experimental contribution. Hertz is best known for his experiments that confirmed the existence of electromagnetic waves, as predicted by James Clerk Maxwell's equations. His work provided the first experimental evidence that light and radio waves are part of the same electromagnetic spectrum. Hertz's discoveries paved the way for the development of wireless communication and a deeper understanding of electromagnetic theory.

- **Work:** *Hertz, H. (1887). "On Very Rapid Electric Oscillations." Annalen der Physik*, 267(7), 421–448.

Hertz, H. (1888). "On the Electromagnetic Effects Produced by Electrical Disturbances in Insulators." Annalen der Physik, 270(10), 155–170.

- **Contributions:**
  - **Experimental Confirmation of Electromagnetic Waves (1887):** Hertz demonstrated the existence of electromagnetic waves through a series of experiments. He constructed a device using a transmitter and a receiver that could generate and detect electromagnetic waves. Hertz observed that these waves had properties similar to light, including reflection, refraction, and interference, providing experimental validation for Maxwell's theoretical predictions.
  - **Discovery of the Photoelectric Effect (1887):** While investigating electromagnetic waves, Hertz observed that ultraviolet light could induce electric sparks between metal electrodes. This phenomenon, later known as the photoelectric effect, played a crucial role in the development of quantum theory and was explained by Albert Einstein in 1905.
  - **Measurement of Wavelength and Velocity of Electromagnetic Waves:** Hertz was able to measure the wavelength and velocity of the electromagnetic waves he produced, confirming that they travel at the speed of light. This finding established that light itself is an electromagnetic wave and solidified the connection between electricity, magnetism, and optics.
  - **Proof of Electromagnetic Wave Properties:** Hertz's experiments demonstrated that electromagnetic waves could be polarized, refracted, and reflected, just like light waves. He showed that electromagnetic radiation was a transverse wave, meaning the oscillation of the electric and magnetic fields occurred perpendicular to the direction of wave propagation.
  - **Development of the Hertzian Oscillator:** Hertz developed the Hertzian oscillator, a device that produced electromagnetic waves by using a high-voltage spark across a small gap. This apparatus allowed Hertz to systematically study the generation and propagation of electromagnetic waves.
  - **Hertzian Waves and Their Detection:** Hertz's experiments resulted in the first production of what became known as "Hertzian waves" (now called radio waves). He used a loop of wire with a gap (receiver) to detect these waves, demonstrating that electromagnetic waves could propagate through free space without a physical medium.

Influence on Later Work: Hertz's experimental verification of electromagnetic waves had a great influence on the field of electromagnetism and the development of modern physics. His work provided the foundation for the invention of radio, television, and other wireless communication technologies. Hertz's confirmation of Maxwell's theory also inspired physicists such as Nikola Tesla to figure out wireless transmission and Guglielmo Marconi to develop the first practical wireless telegraph system.

Hertz's discovery of the photoelectric effect was instrumental in the eventual development of quantum mechanics. Albert Einstein's explanation of the photoelectric effect, for which he received the Nobel Prize in Physics in 1921, was based on the experimental results that Hertz had obtained decades earlier.

> "Electromagnetic waves propagate in a vacuum with a velocity equal to that of light."
>
> —Hertz, H.

5.2 1888 - Nikola Tesla:

Nikola Tesla (1856–1943) was a Serbian-American inventor, electrical engineer, and physicist known for his pioneering work in electromagnetism and electrical engineering. Tesla made groundbreaking contributions to the development of alternating current (AC) systems, wireless communication, and electromagnetism. His inventions and theoretical insights provided the foundation for modern power transmission, radio technology, and the understanding of electromagnetic waves.

- **Work:** Tesla, N. (1888). *"A New System of Alternating Current Motors and Transformers." Transactions of the American Institute of Electrical Engineers*, 5, 308–324.

  Tesla, N. (1891). *"Experiments with Alternate Currents of Very High Frequency and Their Application to Methods of Artificial Illumination." Proceedings of the American Institute of Electrical Engineers*, 8, 317–350.

- **Contributions:**

  - **Development of Alternating Current (AC) Power Systems (1888):** Tesla's invention of the AC induction motor and transformer made it possible to generate and transmit alternating current over long distances. This innovation led to the adoption of AC power over direct current (DC) power, as championed by Thomas Edison, due to its efficiency and ability to be transmitted over vast distances with minimal power loss.

  - **The Polyphase AC System:** Tesla introduced the concept of polyphase AC systems, which use multiple phases of electrical current to produce a rotating magnetic field. This technology is the basis for modern electric power generation and distribution systems. His development of the polyphase motor and transformer revolutionized power transmission, enabling the widespread adoption of AC power.

  - **High-Frequency and High-Voltage Experiments:** Tesla conducted experiments with high-frequency currents and high voltages, which led to the invention of the Tesla coil. The Tesla coil is used to produce high-voltage, low-current, high-frequency alternating current electricity. These experiments contributed to the understanding of electrical resonance and the behavior of electromagnetic waves.

  - **Wireless Transmission of Electricity:** Tesla envisioned and experimented with wireless transmission of electrical power, which he demonstrated using his Tesla coil and other apparatus. He believed that it would be possible to transmit power wirelessly over long distances, and he built the Wardenclyffe Tower in New York as a prototype for his global wireless power transmission system.

- **Electromagnetic Wave Propagation:** Tesla's work on the wireless transmission of energy anticipated many aspects of modern electromagnetic wave theory. He demonstrated the transmission and reception of radio waves before Guglielmo Marconi's work, and his understanding of resonance and electromagnetic waves contributed to the development of early radio technologies.

- **Contribution to Electromagnetic Theory:** Tesla's experiments with alternating currents, transformers, and wireless energy transmission expanded the understanding of electromagnetic fields and their applications. He proposed that electromagnetic waves could be used for wireless communication and energy transfer, theories that later became the basis for radio and microwave technologies.

Influence on Later Work: Tesla's inventions had a profound impact on the development of electrical power systems. His development of AC power transmission enabled the widespread use of electricity for industrial and domestic applications. Tesla's work also influenced the early development of radio technology, as his experiments with wireless transmission and resonance principles were critical in establishing the feasibility of radio communication.

Tesla's work on high-frequency currents and electromagnetic fields inspired future research in the field of radio waves, wireless communication, and electromagnetic resonance. His inventions, such as the Tesla coil, continue to be used in educational demonstrations and are foundational to understanding electromagnetism. His work is still studied and revered for its visionary impact on modern technology and science.

"Through the wireless transmission of electrical energy, we can create an interconnected world where power is freely available without the need for wires."

—Tesla, N.

5.3 1901 - Guglielmo Marconi:

Guglielmo Marconi (1874–1937) was an Italian inventor and electrical engineer who is credited with pioneering long-distance radio transmission and establishing the practical use of radio waves for wireless communication. His experiments and inventions made significant contributions to the field of electromagnetism, particularly in the area of electromagnetic wave propagation and the development of radio technology. Marconi's work laid the foundation for modern telecommunications and broadcasting.

- **Work:** Marconi, G. (1901). *"Experiments in Wireless Telegraphy."* Proceedings of the Royal Society of London, 68(445-453), 344-355.

  Marconi, G. (1902). *"On the Transatlantic Wireless Signal."* Proceedings of the Royal Society of London, 70(460-465), 43-47.

- **Contributions:**

- **Development of the First Practical Radio Transmission System (1895–1897):** Marconi improved on existing technology by developing a practical system that could transmit Morse code signals over long distances without wires. He used a spark-gap transmitter and a coherer receiver to send and receive electromagnetic signals. This system was able to transmit signals over several kilometers, a significant achievement at the time.

- **Invention of the Marconi Antenna (1895):** Marconi designed a vertical antenna system that increased the range of radio transmission. His antenna design helped to reduce the attenuation of signals over long distances and improved the transmission and reception of electromagnetic waves.

- **First Transatlantic Wireless Communication (1901):** Marconi achieved the first transatlantic wireless transmission on December 12, 1901, when he successfully sent the letter "S" in Morse code from Poldhu, Cornwall, in England to St. John's, Newfoundland, in Canada. This historic achievement proved that radio waves could travel over the curvature of the Earth and established the feasibility of long-distance wireless communication.

- **Development of Shortwave Radio and Long-Distance Transmission:** Marconi's research into shortwave and long-wave radio frequencies led to the discovery that certain frequencies could reflect off the ionosphere, enabling even longer-distance communication. This discovery was instrumental in the development of global radio communications and broadcasting.

- **Establishment of Wireless Telegraphy Networks:** Marconi's work led to the creation of the first international wireless telegraphy service. He established wireless communication stations across Europe and the United States, providing a new means of transmitting information over large distances, which had significant applications in maritime navigation and emergency communication.

- **Electromagnetic Wave Propagation and Resonance:** Marconi's experiments contributed to the understanding of electromagnetic wave propagation and resonance, as he demonstrated that radio waves could be focused and directed using specific configurations of antennas and receivers. His work helped validate and expand upon the theoretical predictions made by earlier scientists such as James Clerk Maxwell and Heinrich Hertz.

Influence on Later Work: Marconi's success in developing long-distance radio transmission had a transformative impact on the fields of communication technology. It influenced further development of radio broadcasting, radar, and early forms of wireless communication.

Marconi's contributions also influenced the development of television and satellite communication.

Marconi's work on radio wave propagation and his establishment of the first commercial radio telegraph service provided the foundation for the modern telecommunications industry. His achievements were recognized with numerous awards, including the Nobel Prize in Physics in 1909, which he shared with Karl Ferdinand Braun for their contributions to the development of wireless telegraphy. His research and practical achievements continue to influence modern wireless communication.

"Through the medium of electric waves, it is possible to communicate over vast distances without the necessity of wires."

—Marconi, G.

5.4 Summary of the 19th Century Contributions

The 19th century was a period of profound scientific and technological advancement, which led to transformative innovations in electrical engineering and communication. It saw the development of a unified theory that integrated electricity and magnetism into a single framework, leading to revolutionary technological advancements and a deeper understanding of the physical world.

- **Unification of Theories:** The 19th century saw the unification of electricity and magnetism into a coherent theory through Maxwell's equations, which integrated previously separate concepts into a single framework.

- **Experimental Validation:** Key experiments by Ørsted, Faraday, Hertz, and others validated theoretical predictions and demonstrated practical applications of electromagnetism, leading to significant technological advancements.

- **Technological Advances:** The development of technologies such as the electric motor, telegraphy, radio, and alternating current power systems transformed society and laid the groundwork for the modern electrical and communications infrastructure.

6 20th Century: Quantum Electrodynamics and Relativity

The development of quantum electrodynamics (QED) involved contributions from several scientists, including Paul Dirac, Richard Feynman, Julian Schwinger, and Sin-Itiro Tomonaga. QED describes how light and matter interact and is the first theory where full agreement between quantum mechanics and special relativity is achieved.

Advances in particle physics and the discovery of the electromagnetic interaction's role in the Standard Model of particle physics have furthered our understanding of electromagnetism at the subatomic level.

6.1 1900 - Max Planck

- **Work:** Quantum Theory

- **Contributions:**

    - **Blackbody Radiation:** Planck introduced the concept of quantization of energy to explain blackbody radiation, which led to the birth of quantum theory. His work established the idea that energy is emitted or absorbed in discrete units (quanta), setting the stage for the development of quantum mechanics.

- **Planck's Constant:** He introduced Planck's constant, a fundamental physical constant that relates the energy of a photon to its frequency, playing a crucial role in quantum electrodynamics.

6.2 1905 - Albert Einstein

- **Work:** Photoelectric Effect

- **Contributions:**

  - **Photoelectric Effect:** Einstein explained the photoelectric effect using the quantum theory of light, proposing that light consists of discrete packets of energy called photons. This work provided experimental evidence for quantum theory and was instrumental in the development of quantum mechanics.
  - **Special Relativity:** Einstein's theory of special relativity, published in 1905, demonstrated that electromagnetism is consistent with the principles of relativity. It showed that the speed of light is constant in all inertial frames of reference, profoundly impacting the understanding of space and time.

6.3 1920s - Development of Quantum Electrodynamics (QED)

- **Key Figures:**

  - **Paul Dirac:** Dirac developed the Dirac equation, which describes the behavior of relativistic electrons and predicts the existence of antimatter. His work laid the groundwork for quantum electrodynamics.
  - **Richard Feynman, Julian Schwinger, and Sin-Itiro Tomonaga:** These physicists developed quantum electrodynamics (QED), a quantum field theory that describes how light and matter interact. Their work involved complex mathematical formulations and introduced the concept of Feynman diagrams to visualize interactions between particles.

6.4 1930s - Advances in Particle Physics

- **Work:** The Standard Model

- **Contributions:**

  - **Development of the Standard Model:** The 1930s saw the formulation of the Standard Model of particle physics, which describes the electromagnetic, weak, and strong nuclear forces and classifies all known elementary particles. This model incorporates QED and explains a wide range of particle interactions.

6.5 1940s - Development of Microwave Technology

- **Work:** Radar and Microwave Technologies

- **Contributions:**

  - **Radar Development:** During World War II, radar technology was developed and refined, using electromagnetic waves to detect and locate objects at a distance. This technology became crucial for military and civilian applications.
  - **Microwave Ovens:** The invention of the microwave oven in the late 1940s by Percy Spencer, who discovered that microwaves could be used to heat food, demonstrated a practical application of electromagnetic waves.

6.6 1950s - Advancements in Electronics

- **Work:** Solid-State Electronics

- **Contributions:**

  - **Transistors:** The invention of the transistor in 1947 by John Bardeen, William Shockley, and Walter Brattain revolutionized electronics by providing a reliable means of amplification and switching. This invention enabled the development of modern electronic devices, including computers and smartphones.
  - **Integrated Circuits:** The development of integrated circuits (ICs) in the 1950s and 1960s allowed for the miniaturization of electronic components, leading to the proliferation of consumer electronics and computing technology.

6.7 1960s - Quantum Field Theory and Unification

- **Work:** Quantum Field Theory

- **Contributions:**

  - **Quantum Electrodynamics (QED):** QED was further developed, providing an accurate description of the interactions between photons and charged particles. This theory has been tested with remarkable precision and has become a cornerstone of modern physics.
  - **Electroweak Theory:** The electroweak theory, developed by Sheldon Glashow, Abdus Salam, and Steven Weinberg, unified the electromagnetic and weak nuclear forces into a single theoretical framework. This work earned them the Nobel Prize in 1979 and contributed to the development of the Standard Model.

6.8 1970s - The Standard Model and Experimental Confirmation

- **Work:** Experimental Confirmation of the Standard Model

- **Contributions:**

- **Particle Accelerators:** The use of particle accelerators, such as the Large Hadron Collider (LHC), allowed for experimental confirmation of various predictions of the Standard Model, including the discovery of the W and Z bosons, which mediate the weak force.

6.9 1980s - Advancements in Telecommunications

- **Work:** Fiber Optics and Wireless Communication

- **Contributions:**

    - **Fiber Optics:** The development of fiber optic technology revolutionized telecommunications by allowing high-speed data transmission over long distances with minimal loss. This technology relies on the principle of total internal reflection of light in optical fibers.
    - **Wireless Communication:** Advances in wireless communication technology, including cellular networks and satellite communications, transformed global communication by enabling wireless data transmission and connectivity.

6.10 1990s - Emergence of the Internet and Modern Computing

- **Work:** Internet and Computing Technology

- **Contributions:**

    - **The Internet:** The widespread adoption of the Internet in the 1990s facilitated global communication and information sharing. The Internet relies on electromagnetic signals for data transmission over various communication channels.
    - **Computing Power:** Advances in computing technology, including the development of microprocessors and high-speed digital circuits, enabled the growth of personal computers, the World Wide Web, and digital technologies.

6.11 Summary of the 20th Century Contributions

- **Quantum Electrodynamics (QED):** The development of QED and its incorporation into the Standard Model provided a comprehensive theoretical framework for understanding electromagnetic interactions at the quantum level.

- **Technological Innovations:** The 20th century saw the invention and refinement of technologies such as transistors, integrated circuits, radar, microwave ovens, fiber optics, and wireless communication, all of which are grounded in electromagnetic principles.

- **Theoretical Advances:** The unification of forces in the Standard Model, the development of electroweak theory, and the experimental confirmation of theoretical predictions advanced the understanding of fundamental interactions.

Overall, the 20th century was a period of rapid progress in both theoretical and applied electromagnetism. The advancements made during this time have had a profound impact on technology, communication, and our understanding of the fundamental forces of nature.

7 21st Century

7.1 Terahertz Technology Development (2000s)

- **Terahertz Wave Experiments:** Researchers developed techniques to generate and detect terahertz waves, including terahertz time-domain spectroscopy and imaging. These experiments explore applications in material science, security, and medical imaging.

7.2 Quantum Computing Experiments (2000s-Present)

- **Superconducting Qubits:** Experiments with superconducting circuits to create and manipulate qubits have been central to quantum computing research. These experiments involve precise control of electromagnetic fields to achieve quantum superposition and entanglement.

7.3 Metamaterials Research (2000s-Present)

- **Metamaterial Experiments:** Researchers have created and tested metamaterials with engineered electromagnetic properties. Experiments include studying negative refractive index materials and designing devices for applications like cloaking and superlenses.

7.4 Quantum Communication Experiments (2010s-Present)

- **Quantum Key Distribution (QKD):** Experiments in quantum communication involve the use of photons for secure data transmission. Researchers have developed and tested QKD protocols and quantum networks to ensure secure communication based on quantum principles.

7.5 Advanced Imaging Techniques (2010s-Present)

- **Super-Resolution Microscopy:** Experiments in super-resolution microscopy use advanced optical techniques to achieve imaging resolutions beyond the diffraction limit. These techniques utilize electromagnetic waves to observe biological and nanoscale structures with high precision.

7.6 Wireless Power Transfer (2010s-Present)

- **Wireless Charging Systems:** Experiments with inductive and resonant wireless power transfer have led to the development of practical wireless charging systems for consumer electronics and electric vehicles. Researchers focus on improving efficiency and range.

7.7 Summary of 21st Century Contribution

These experiments represent significant milestones in the exploration of electromagnetism, ranging from early investigations of static electricity and magnetism to modern advances in quantum communication and terahertz technology. Each experiment contributed to the expanding understanding and application of electromagnetic principles, shaping the technological landscape and scientific knowledge of the field.

8 Conclusion

The history of electromagnetism spans over two millennia, from ancient Greek observations of static electricity to the theories of quantum electrodynamics in the 21st century. The understanding has evolved from ancient observations to a sophisticated theoretical framework, underlying the development of modern electrical and electronic technology.

Pioneers like Thales, Gilbert, Franklin, Faraday, Maxwell, Hertz, Tesla contributed to the technological advancements that have shaped the modern world, from electric power generation and radio communication to the development of quantum mechanics.

Each discovery has built on the last, contributing to the comprehensive electromagnetic theory that is central to contemporary physics and engineering.

There are still many unanswered questions on electromagnetism left for us to understand in each of the phenomena we studied through our historical tour. We are still waiting for something new to come out of the same ancient observations.

Leave a review for further improvement

Figure 1: **Every review will help us to understand more.**

www.ingramcontent.com/pod-product-compliance
Lightning Source LLC
Chambersburg PA
CBHW081021240526
45471CB00018B/3924